Catharine Waugh McCulloch

**The Legal Status of Mother and Child**

Catharine Waugh McCulloch

**The Legal Status of Mother and Child**

ISBN/EAN: 9783744648431

Printed in Europe, USA, Canada, Australia, Japan

Cover: Foto ©berggeist007 / pixelio.de

More available books at **www.hansebooks.com**

# MR. LEX

OR

The Legal Status or Mother and Child

BY

Catharine Waugh McCulloch

Chicago : New York : Toronto

Fleming H. Revell Company

1899

# TO MOTHERS,

SOME OF WHOM NOW SUFFER THROUGH UNJUST LAWS;

# TO FATHERS,

MANY OF WHOM ARE KINDER THAN THE LAWS AND ARE PROBABLY IGNORANT OF OR INDIFFERENT TO POSSIBLE INJUSTICE;

# TO LEGISLATORS,

WHO HAVE THE POWER TO CHANGE LAWS.

# PREFACE.

In the following pages the form of fiction is used in order to place more vividly before the reader the injustice which results from laws which make fathers sole guardians and custodians of their children. The incidents related are all paralleled in real life. The court decisions described are based on statutes and actual decisions of courts. The authority for all statements of law can be found in the table of citations appended. The numerals in parentheses through the text refer to the citations in the table bearing the same number.

Lack of space prevents the presentation of authorities from every State, and so only Illinois authorities are cited. But as three-fourths of our States make fathers the sole guardians and custodians of children, and deprive mothers of such authority, the courts in these States, in circumstances similar to those narrated, render similar decisions. This makes the question of parental authority of more than local importance.

These laws originated many years ago under military despotisms, when fathers alone were strong, educated, responsible, possessed of property, and skilled in war. The welfare of the children, the good of the State, seemed to require that the ignorant, dependent, helpless mothers should not be invested with authority over their children.

But the conditions which gave rise to these laws no longer prevail. Mothers are today responsible, cultivated, discreet, experienced, and, under the changed conditions of society, are entirely capable of sharing with fathers in the guardianship and control of their children. The laws should be so changed as to allow mothers such power.

Other laws relating to women have changed, as law-makers have slowly recognized women's increasing ability and power. All our States now allow women to devise and bequeath their own property; three-fifths of our States allow them to possess, manage, and convey their own real and personal property; nearly all the States allow them equal rights under the divorce laws; four States grant them complete rights of suffrage; one grants them municipal suffrage and twenty-three, school suffrage; some States have so far recognized mothers' obligations to the family as to make them liable for family expenses; and about one-fourth of the States recognize in some way the authority of mothers over their children.

It must not be forgotten that, though women have petitioned for more equitable laws, these great changes have only become possible through men's generosity, when they were persuaded that the good of women, of children, of the home and of the State demanded improved conditions.

So there is ground for believing that men will continue to make alterations in laws as soon as they are convinced of the necessity.   Laws should be changed so that fathers and mothers may be joint guardians and custodians of their children with equal responsibility·and authority.

If this book shall in any degree help mothers to a clear vision of their own responsibilities; if it shall convince fathers that mothers, deprived of power, can never do their full duty toward their children; or if it shall induce legislators to change unjust laws, its·purpose will be accomplished.

CATHARINE WAUGH⊐McCULLOCH.
*Chicago, January, 1899.*

# MR. LEX.

# MR. LEX.

Mr. and Mrs. Lex and their five children lived in a small flat in Lake View, an outlying section of Chicago, about two blocks away from the grocery store which Mr. Lex managed. He had not always been a grocer, but prided himself on having an education superior to his employment and surroundings. He had been educated for the profession of law, and at the time of his admission to the bar had been considered a clever, bright young fellow but somewhat conceited and overbearing. His wife had been very proud of him as she fondly pictured his future successes. But, after ten years of waiting for clients, he at last decided to give it up. Either his nervous, irritable manner, or lack of thoroughness, or one or two great blunders, or lack of tact, or all combined, had conspired to bring him few clients and little income.

During these early years of waiting for success, his wife had devoted to family expenses the money which her father had bequeathed to her. Little children had been coming into the home, and increasing needs had greatly diminished her inheritance. It had not been large in the first place; for her father had devised the bulk of his property to his son, his theory being that, as husbands support wives, daughters need but little. Mrs. Lex's mother protested against this injustice, and urged that the property she helped him to earn should be divided equally between her son and daughter. But, as is generally the case, the title to everything was in the husband's name; so he was not obliged to heed his wife's protests, nor did he.

So Mrs. Lex learned early, as her first lesson on the legal status of mother and child, that her own mother, because she had nothing in her name, could not give or bequeath to her any of her own savings or earnings, although the statutes of Illinois even then provided that married women might control their earnings and convey (1) or bequeath them (2). She also learned that, although our laws concerning inheritances make no difference as to sex when the father dies without a will, yet he can make a will disinheriting his child.

Of her inheritance, a little grocery store was the only thing now left, and Mrs. Lex had finally induced her husband to go into that, since ten years of the practice of law with few clients and the exhausting of her own funds had brought them to poverty. So he descended, as he called it, into a grocery store; but he never forgot that he had learned something about law, that he was a lawyer, and that he had sworn to support the constitutions of the United States and of the State of Illinois and faithfully to discharge the duties of the office of attorney and counsellor at law. His subsequent conduct showed that he kept his vow even when it necessitated breaking others holier and higher.

The oldest son, John, helped in the

store. He was not quite sixteen years old. Mary was fourteen, Jennie was eleven, Rob was nine and Daisy six.

Mrs. Lex, though a pretty girl at eighteen, when she married, was now, at thirty-five, a faded, careworn woman with much of her early hope gone. She had economized and worked hard, caring for the family, with no help except for a few weeks when her children were very young. To sew, wash, bake, and clean for a family increasing from two to seven in ten years had been no small task, but she could have endured these physical burdens cheerfully if her husband had been as tender as he was when they were first married. Now he grumbled, he argued, he scolded; when he was asked for money he always objected, and declared that he furnished the family all the law required. As to food, that could be got out of the store, and as to clothes, he chose what he thought best. His wife gently hinted that with the same money she could do better, but he told her the law never provided that she should dictate in what manner his money should be spent by selecting things herself. So he put Rob into long trousers when he was nine, though Rob complained that he could not run well. He chose green, checked green, for John's last suit of clothes, the one color above all others that John abominated and that made him look ghastly. When Mrs. Lex said that John would enjoy better a suit of brown or gray he told her that it was his place to choose colors. She keeping house at home, with no salary, had not a cent of money. But he was getting some money now from the customers at the store and could be more arrogant than when the money came through his wife.

When he took Mary down street he bought a pink felt hat trimmed with yellow chrysanthemums and a carmine bird. He purchased also a purple jacket and a blue dress. Everything was of durable material, he said, and he liked a variety of color. When Mrs. Lex beheld pretty Mary in this array, she again urged that she could better select the clothes, and he again told her they must submit to his selections.

A neighbor had once informed her that she could buy what she needed at any store, have it charged to her husband, and he would be obliged to pay for the article. So she bought a neat gray jacket for Jennie, who had had no winter coat for three years, and had it charged to her husband.

When the bill came Mr. Lex refused to pay it, and so he was sued, the dry-goods firm claiming that this was a family expense for which he, a parent, was legally liable. Mr. Lex defended his own suit, asserting that this was not a family nor a necessary expense. Jennie could wear the plaid green-and-red shawl she had worn for the past three years. Mrs. Lex whispered to him that she herself needed it to wear under the thin, old one which had been her only wrap for years, but he hushed her into silence.

He said, "This jacket is not needed for the family; 'it in no way conduces to the welfare of the family, generally.' (3) It is 'for personal adornment, largely to gratify vanity; and though it may be incidentally worn in the family, its primary and important use is for display in general, society.' Therefore, in accordance with the decision of the Supreme Court (3) this could not be a family expense for which I would be liable." The lawyer for the firm objected, "This decision quoted from was made in reference to a diamond ring, a purchase very different from that of a child's cheap jacket."

Mr. Lex replied: "The principle is the same, and should apply here. Besides, I gave my wife no authority to make this purchase and, in fact, had really forbidden her. The merchant did not show any authority; and as 'a tradesman who sells goods to the wife upon credit of her husband takes the burden of proof of showing such authority' (4) I am not liable. Only I, the father, the legal guardian of the children, should decide upon such matters." The Justice was a stickler for legal points, and so decided in Mr. Lex's favor.

But the dry-goods firm now turned on Mrs. Lex and threatened her with arrest for obtaining the cloak under false pretenses. Jennie, who was in tears, begged them not to arrest her mother, and became so frightened that she stripped off the modest jacket and handed it over to a representative of the firm.

Mrs. Lex did not try to buy any more clothes, and her husband did not let her forget that his choice must prevail. She thus had another lesson concerning the legal status of mother and child. She discovered that it was very different from that of father and child.

Mr. Lex devised unusual and humiliating punishments for the children, and grew angry when his wife expostulated. He showed her that the Supreme Court of Illinois had said in relation to the punishment of a child, "The authority to govern must rest in some one, and the law has placed that power in the hands of the father as the head of the family." (5)   So Mrs. Lex tried to submit.

About this time the father began to be interested in the spiritual welfare of his children, and directed them to go no longer to the neighboring Presbyterian church, which they had always attended, but to go to the Adventist's church two miles away.   They all objected, but to no avail.

He explained: "Choosing the church where you worship is the privilege of your legal guardian, which position I alone hold.   No one else has the right to interfere."

Who dared to contradict this ex-lawyer? Besides, he was right. So, when they went, they had to go to the church of his choosing, which was no light cross, considering the clothing they wore.

He also insisted on periodic family dosings of sulphur and molasses, and in spite of his wife's objections prescribed for the children certain injurious articles of diet, declaring that the law laid on him the duty of nurturing his children, and that these things were needed by the children for their nurture. Mrs. Lex wanted the children to wear flannel in winter, but he said they must be toughened.

Little Daisy had always been delicate, and this winter she coughed. Rubbers might have protected her feet, but the father would not buy rubbers, "because they make the feet perspire." One cold, damp Sunday he took her walking without rubbers and of course without flannels. That night she had croup. Mrs. Lex begged her husband to call a physician, but he said Daisy did not need a doctor, and he as her legal guardian would not allow one to be called.

So Mrs. Lex dared not call one; and besides she had no money and she remembered the cloak episode. She tried all the home remedies but with little effect. By the next morning Mr. Lex made up his mind to call a doctor, who appeared only in time to see little Daisy gasping her last.

You can imagine the grief of the mother, and as she was human she did reproach the father and told him he was responsible. He said, "It is a mysterious dispensation of Providence. I have often been sicker myself and pulled through, and I had expected Daisy would. If I had called a physician quicker she might have died sooner Physicians use poisons half the time, anyway. It will always be a comfort to me that I have kept the law and have chosen for this child what I considered best."

Cousin Jane, a widow, had a cemetery lot at Graceland, and offered to give them a portion in which to lay the little body. Mrs. Lex thankfully accepted her generosity, for this would be so near she could occasionally visit the spot. But Mr. Lex was astonished at his wife's presumption. He said, "Do you suppose the law gives you the right to select the burial place for my child? I am the guardian of that child, alive or dead, and I shall take her to Oakwoods."

The mother pleaded that she and the children could not visit the grave there, and begged that Cousin Jane's offer be accepted. She even went to see the Justice of the Peace and asked why she could not have the body buried as she wished, near home. The Justice sympathized with her, but said her husband had the right to bury the body of the child where he chose, as he was the legal guardian of the child's person. In what a bitter school was she learning about the legal status of mother and child!

Her big boy John was the next one about whom she worried. He was nearing sixteen, and was shooting up into a tall lad. He was busy all the time in the store, and Mrs. Lex urged her husband to give him less of the heavy work unsuitable for a growing boy. She thought he ought not to lift heavy barrels of flour and sugar. Mr. Lex again reminded her that such matters were beyond her control, for he, the father, had the authority to designate his son's duties. One day John grew suddenly pale when lifting one of these barrels, and his mother, who was at the store getting the day's supply of groceries, flew to his side and asked the trouble. "Something seems to have given way," he faintly said, and Dr. Jones, who happened in at the time, examined him and said that the severe strain had injured John internally so that he could never again do heavy manual labor.

It is not necessary to describe their respective feelings nor the father's legal quotations nor the mother's despair. A few weeks later, when John was better, though he never could be cured, his mother defied the father sufficiently to go to a small dry-goods store, where she secured for John a place as cashier at five dollars a week. The business manager was an old friend.

There was grumbling from the father, but he relented sufficiently on payday to collect John's wages. The proprietor had intended to pay John, but the father explained that John was only a minor and that, under the law, his wages belonged to his father. So for some weeks he collected five dollars per week, justified by the consciousness that he was living up to the law. Later the dry-goods firm made a rule not to pay wages until a month had passed, and so twenty dollars of John's unpaid wages accumulated, when the lad was dismissed to lessen expenses.

Mr. Lex was having a slight attack of rheumatism which had kept him at home several days, and Mrs. Lex did not want to arouse his unusually irritable temper. So she asked the dry-goods firm for the twenty dollars John had earned, but was refused. The family needed this money sadly, and in her perplexity she went in to see the nearest lawyer, who said she should sue the firm for the money. She began suit. The matter came before the same Justice who presided at the cloak trial. It was proved that the dry-goods firm owed twenty dollars for the services rendered by John, and that she was his mother.

The attorney asked her, "Are you widowed or divorced or deserted?" She said, "No, my husband is at our home a few blocks distant."

Then the attorney argued as follows: "The father of this minor, his guardian, is the only one entitled to collect John's wages. (6) Even a step-father has a right to a minor's services before the mother. (7) It is only when the father is divorced or dead (8) or incapacitated by insanity (9) or has abandoned his family (10) that the mother can take his position."

The Justice, in the face of this correct legal statement, could do no less than decide that, as Mrs. Lex was not the boy's guardian, she was not entitled to his wages and must pay the costs of the suit. He added kindly, however, that he was sorry for her but that law, not pity, must guide him in his decisions.

This was another lesson in relation to the legal status of mother and child.

The condition of her daughter Mary now began to distress the mother. Mary was troubled, and was very different from the happy girl of a few months ago, but she gave no reason for the change. Young as she was, she had frequently helped in the store; and there, away from her mother's watchful eye, an evil man had often talked and laughed with Mary and had then walked home with her after the store closed at eight in the evening.

It was the old story. He pretended to love her, and, as she had none too much love from her irritable father and weary mother, to her affectionate heart the love proffered by this handsome man of twice her years was very welcome. She had never been warned or instructed by the unthinking mother, and it was an easy task for the skillful villain to make her his victim; then he suddenly ceased talking of love and ceased coming to see her.

At last Mary told her mother. Mrs. Lex was agonized. She could scarcely believe that Mary was approaching maternity. Mary, not yet fifteen, still in short dresses, with two braids down her back! Mary, only yesterday playing with dolls, now under the shadow of the gravest responsibility and the most cruel shame!

Mrs. Lex dreaded to tell her husband, but, knowing his profound respect for law, she felt sure there must be some punishment for this great wrong. She went to the Criminal Court Building to see the State's Attorney, and asked if he could not send the villain to the penitentiary. He asked Mary's age at the time of the wrong. She had just passed the anniversary of her fourteenth birthday.

"That's bad," he said. "Girls over fourteen are considered by our lawmakers old enough to consent to their own ruin. (11) Villains know this, take advantage of the law and the girls' ignorance, and as a result our hospitals have many mothers under fifteen. But did he use violence? Was this done 'forcibly and against her will'? (12) If so, the wrong would be called rape, and he might be sent to Joliet to the penitentiary for a year or two, as he could if she had been under fourteen."

Mrs. Lex replied: "He did not use

physical force but the persuasion of a lying, treacherous tongue."

The State's Attorney then read to her out of an old law book that a woman has hands with which to beat, nails with which to scratch, teeth with which to bite, feet with which to kick, and a voice with which to make an outcry, and should she fail to use all these natural weapons of defense to the utmost, she would not be presumed to have objected seriously to her own ruin but to have consented, and the wrong would not be a crime, not a penitentiary offense. Alas, Mary had not been taught to fight; she was no Amazon; she was only a loving, gentle, ladylike little girl.

The State's Attorney was a busy man, and Mrs. Lex was but one of many mothers with such tales; but he sympathized with her, and took time to advise her that the man would be liable in an action in bastardy and, if proved guilty, might be compelled to pay as much as five hundred and fifty dollars during the next ten years. (13)

But he added that sometimes jurors did not allow five hundred and fifty dollars, as they were men and might be expected to have a little sympathy for an erring brother man. Then two hundred or three hundred dollars was all that would be paid. These payments were not supposed to be for punishment but only for the support of the child. If the child died, the payments would be stopped. (14) The sum of one hundred dollars would be paid the first year and fifty dollars each succeeding year; or if the whole sum was less than five hundred and fifty dollars, the payments would be proportionately smaller. This would be the father's only contribution to the support of the child. (15) "No father," continued the State's Attorney, "is obliged to support an illegitimate child except to the extent of this allowance. (16) An illegitimate child cannot even inherit property from his father. (61) The rest of the funds necessary to support the illegitimate child must come from the mother, and he can inherit property only through the mother. If the child should be an idiot or infirm of body and thus be too helpless to support himself, the mother's liability for his support would continue after he became of age up to the time of his death, when the funeral expenses must be paid by her."

"However, Mrs. Lex," said the State's Attorney, "you might get more money out of the man if you should sue him for damages resulting from loss of the daughter's services. Then you need not trouble with a proceeding in bastardy." The State's Attorney recommended that, if she began such a suit, she should employ young Smith, an honest, hardworking lawyer, who would not be bought off.

Mrs. Lex said that if the man could not be sent to the penitentiary it was a shame, but that no money could right the wrong done Mary, and she would scorn to ask for it. Later she remembered his advice, but now she awaited the future in despair. In a few weeks Mr. Lex found out his daughter's condition and his anger was frightful. He stormed and raved and turned Mary out on the street; and when Mrs. Lex begged to have her stay, he locked Mrs. Lex in her room.

He told her no law compelled him to keep an obnoxious child under his own roof, especially when she was past the age of twelve years. (17) He might be made liable for her support, perhaps, but he was not obliged to keep her at home.

Poor Mary tottered along the street, scarcely seeing where she was, and went right into the arms of Cousin Jane, she who had offered the cemetery lot at the time of the death of Daisy. Cousin Jane had her own little cottage on a side street about two blocks away, where for years she had done plain sewing for several wealthy families near Lincoln Park. She took Mary home with her, notified Mrs. Lex next morning as to Mary's whereabouts, and said she would give the poor girl a shelter. Mrs. Lex was very grateful. She was troubled, however, as to how the expenses could be borne during Mary's approaching sickness. Cousin Jane really could not afford to do everything.

Then Mrs. Lex remembered what the State's Attorney had said about damages. If she had a few hundred dollars, she could pay Cousin Jane for actual outlays and be indebted to her only for her great kindness of heart. So Cousin Jane was authorized to employ Lawyer Smith, who was recommended by the State's Attorney, and he sued the seducer for damages in the name of Mrs. Lex, the mother.

We pass briefly over the occurrences of the next year. Mary's baby was born before she was fifteen, and Mary stayed on with Cousin Jane, helping with the sewing. Mr. Lex never saw or asked about her, and Mrs. Lex said nothing. When he heard about the suit he told Mrs. Lex she would be sorry for it, and she was, later.

Young Smith rushed the case to trial and, to the astonishment of almost every one, got a verdict for twelve hundred

dollars. Now Cousin Jane could be
paid, they thought. But the angry de-
fendant insisted upon appealing the
case, his attorney contending that twelve
hundred dollars was a preposterous
amount; as Justice Gummere, of New
Jersey, had held that one dollar would be
sufficiently large damages for the loss
of a child, and Judge Ferris, of Ohio,
had held that children were not assets
but liabilities.

In an unusually short time the Appellate Court decision came. In the opinion, the Court said: "This sort of an
action is based on the theory that
Mary's ability to labor has been dam-
aged through the seduction, and this is a
wrong done to the one who is entitled
to her services or wages, the father and
not the mother (18). The father has
the exclusive right of action. Not even
the daughter herself could maintain an
action for her own seduction. Even
when a father has turned his daughter
out on account of pregnancy, he is held
to be the proper person to bring suit.
(19) From the year 1842 (20) our

highest courts have upheld the right of
the father to sue for damages resulting
from the loss of his daughter's services
through seduction (21). The Supreme
Court has affirmed the right of a near
relative of an orphan, the right of a
master or a guardian (22), but has never yet sanctioned a mother's bringing
such action, and never would if it did
not affirmatively appear that there was
no father entitled to the daughter's services. A case exactly similar came before this court (23) in which the court
said: 'It was therefore incumbent upon
plaintiff in the court below to prove that,
at the time of the alleged seduction, she
was entitled to or had the right to com-
mand her daughter's services to her
own use. Ball vs. Bruce, 21 Ill., 161;
Doyle vs. Jessup, 29 Ill., 460. This we
think she failed to do. The presumption is that the daughter was born in
lawful wedlock and that her father is
still alive. If so, the right of action is
in the father and not in the mother.
The right of the mother to the custody
of her minor child does not arise during

the lifetime of the father unless so ordered by a court in a proper case.' Hobson vs. Fullerton, 4 Ill. App., 284. The decision in this Hobson case is decisive in the case at bar."

The Court also said: "Sympathy might lead us to decide differently, but we are bound by law and precedent. 'It has long been considered a standing reproach to the common law that it furnished no means to punish the seducer of female innocence and virtue, except through the fiction of supposing the daughter was a servant to her parent and that in consequence of her seduction the parent had lost some of her services as a menial. It is high time this reproach should be wiped out.' Anderson vs. Ryan, 8 Ill., 588. Some States have made seduction a criminal offense, and have thus shown a proper horror of such great wrong. But in Illinois there seems little redress for such evil. It is considered merely a misdemeanor, a misbehavior, a breach of manners. Even the bastardy act, which provides for the payment of money for the support of the child, is not the imposition of a penalty for any immoral or illegal act but a civil remedy." (24).

The Judges evidently recognized the injustice of such a decision even if convinced of its legal correctness, for in conclusion they said, "Judges can merely enforce laws as they are. Until the Legislature of Illinois makes some change, such wrongs will often go unpunished. There is now, in this case, no punishment for the wrongdoer. He cannot be punished for rape, for the wrong was not done forcibly or against her will, and Mary was over fourteen. Neither she nor her mother has a right to bring this suit, and the father refuses. It is now too late to begin proceedings in bastardy. This is a case where there is clearly a wrong but no remedy."

Mr. Lex had been very angry that his wife had anything to do with "that abandoned girl," as he called Mary; and he now rejoiced in her final defeat because it must impress on her forcibly her own powerlessness and his full authority over their children.

The State's Attorney wrote, asking her to call at his office. When he saw her he expressed great regret that the case had terminated so unsuccessfully. He said, "If only you had told me your husband was living I should never have advised you to begin proceedings, but I thought you were a widow. I have watched this matter with great interest, and now almost feel as though I were responsible for your failure and the heavy legal expenses incurred. I, too, have daughters, and it cuts me to the heart to think of what you and your little girl have suffered. I have here one hundred dollars which I want you to accept for legal expenses or other debts incurred through Mary's trouble. Do not count it a gift but the collection of a fine which I have imposed on myself for belonging to a sex which has wrought such misery."

The State's Attorney cleared his throat, turned his back, and asked his clerk to show her out before she could collect words in which to express her gratitude at this gift. He had so placed the matter before her that she could accept the money without humiliation.

She took the money to young Smith to apply on his bill, but he said, "I like to try cases just for the experience it gives me; and I have learned so much during the pendency of the suit that I feel that you owe me nothing. I have settled for the printing of abstract and briefs for the Appellate Court and for the costs, just for the experience too, and now here is a receipt in full."

Mrs. Lex said he was too generous, but he said, "It is only a form of selfishness. Perhaps my little baby girl may need a friend some day, and this method will be surer to secure friends for her than accumulating government bonds. I wish you to feel that you and Mary ought to be ready to help my daughter or any other troubled girl if in need."

Mrs. Lex said she would never forget his generosity nor that of the State's Attorney, for to her they would ever be the ideal, nineteenth century, chivalrous gentlemen. This made young Smith look so red and fierce that the office boy, at a distance, thought he was angry; but Smith controlled himself, and the office boy heard nothing violent.

Mrs. Lex's sad heart was comforted by the sympathy of these two noble men, who tried to show her kindness because the law showed no justice. She flew about with the money to physician, nurse, and druggist, and every bill was paid except the debt they owed Cousin Jane. That would never be paid now, for Cousin Jane had died a few days before, and their best friend was gone.

When her will was read it appeared that she had no nearer relatives and that she had devised the tiny cottage to Mary, and had given all her money savings, four hundred dollars, to Mrs. Lex. Young Smith had made the will, and had probably encouraged Cousin Jane to such a disposition of her estate.

Mary had become a skillful needlewoman, and she continued to hold all

Cousin Jane's customers. So she and her baby were sure of food and shelter.

Mrs. Lex was grateful for the four hundred dollars, for she knew the many needs that the coming winter would bring which she could now supply. There would be flannels and warm shoes for every one and soft new clothes for the tiny little stranger she was expecting in her own family; and perhaps she could hire the washing and ironing done a few weeks till she was strong. Mr. Lex asked her for the money, and she refused to give it to him as she had a legal right to do.

But he spread abroad the news of her legacy, and she was requested by Messrs. Snuff & Co. to pay her husband's tobacco bills; by the downtown store to pay for that horrible pink hat of Mary's, bought three years before; by the druggist to pay for medicines bought and used by Mr. Lex, and by various other creditors to pay bills incurred by her husband for family expenses, including a bill for trousers for himself.

She refused to pay and was sued. Every creditor got a judgment against her. It was urged that these items were all family expenses and necessities. She protested that tobacco for her husband was not a family expense nor a necessity. But Snuff & Company called in a dozen men from the street who all swore that tobacco was a necessity and that it was an ordinary family expense. The Justice said: "The wife's liability for family expenses is not limited to necessities, it applies to 'all expenses of the family without qualification as to kind or amount, and does not depend upon the wealth, habits, or social position of the party. The husband is the head of the family. He determines primarily what is needed for it' (25). Mrs. Lex will be liable out of her own funds for the tobacco." (26)

She objected to paying for the medicines, but the Justice told her that the Appellate Court (27) had decided that a wife must pay for her husband's medicines.

She objected to paying for John's green clothes,—so much abhorred,—and said she never selected them. The Justice said it made no difference. "The wife is not expected to choose family supplies. The husband determines primarily what is needed for the family. He may buy and contract debts all in his own name for the support of the family. It is not necessary that his wife's name be known or her consent be given to bind her for goods purchased by him for the family or for an individual member of it. The Appellate Court has twice (28) passed directly on this question. The clothes are family expenses, and she must pay for them."

Her lawyer, young Smith, urged that John's wages, the twenty dollars which had been unpaid for three years, should be set off and allowed against this demand. Surely John's wages ought to pay for his clothes.

But the Justice explained, "While the father lives at home he alone can collect a minor's wages; and as he is not a party to this suit nor claiming this set-off, the set-off will not be allowed. But a special statute (29) makes both mother and father liable for a child's support, and a creditor can sue either one, even the one who is not eligible to receive the minor's wages. This statute is a step in the right direction, recognizing the mother's responsibility in the home, as a partner who must bear her share of the burden of family expense. Perhaps the law should also allow her a partner's right to a share in family funds with which she might pay such expenses. But as it does not at present I can only enforce law as it is, not as it should be."

"In others words," said Smith scornfully, "a mother is eligible to all duty, all burden, but ineligible to receive benefit in the shape of wages and ineligible to direct the expenditure of her own funds."

Smith protested that she ought not to pay for her husband's trousers, as they were of no use to her. But the Justice said it had been frequently held that a wife must pay for her husband's clothes. (25)

Smith argued: "This four hundred dollars is exempt, the statute providing that four hundred dollars worth of property is exempt." But the Justice said, "Only one hundred dollars is exempt for her, a wife. as the three hundred dollars exemption can be claimed only by the head of the family." (30)

Smith protested that Mrs. Lex really was the head of the family, as all the family's funds had come through her and because she owned even the little store in which Mr. Lex carried on the grocery.

But the Justice replied: "Even if the business were the wife's and the husband acted but as her clerk, 'this did not divest him of his legal functions as the head of the family. He has not abdicated nor forfeited his headship. Nor is that question to be determined merely by ascertaining whether he or she contributed the greater sum to the support of the family.' (31) 'There was no notice that any anomalous relations existed in the family constituting the wife the head of the family. Where the wife lives with the husband, he must be regarded as the head of the family.' (32) Our Supreme Court made this statement in a case where the wife and husband and her children by a former husband lived together on the wife's premises, where all the property belonged to her. Although in reality her husband was living with her, still, legally, she was living with him and he was the head. The ownership of property makes little difference as to headship. The parent seems to need no other qualification than that of masculinity to make him head. Mr. Lex has that. He is the head. The rule is that the husband is the head of the family." (33)

"In case of the husband's insanity (34) or death (35), or his abandonment of her (36) the law esteems the wife the head. As Mr. Lex is sane and living and still cleaving to his family, Mrs. Lex cannot be the head. So only one hundred dollars is exempt from the necessity of paying these claims."

So the three hundred dollars was used to pay for the tobacco she abominated, the clothing she never had selected, and the medicine she never had used.

These lessons as to the legal status of mother and child were becoming expensive.

A few weeks after this little Dora was born, and the poor mother had heart and mind too full to think much about law. Besides, another grief had come to her through her daughter Jennie.

Some time before this Mr. Lex had decided that a different education was needed by Jennie that that afforded by the public school, and so he sent her to a certain church school, where she was taught deportment and embroidery and little else. Mrs. Lex begged him not to send Jennie to this school, where the teaching was so opposed to the faith in which the family had been reared.

Mr. Lex thundered out, "It is my duty as that child's guardian to select her school and to decide as to her tuition, and I shall not be deterred by your foolish whims."

So Jennie was less with her mother than formerly and less influenced by her. At last, when she was nearly fourteen years old, she announced that she was soon to be married.

Mrs. Lex told her that she was too young, that the lover was a dissipated man, and that it could never be allowed. Mr. Lex, hearing the discussion, was so annoyed at his wife's presuming to act as though she had authority over the girl, that he declared: "I am the one to give or refuse consent to my child's marriage (37). If you oppose it, that is a sure sign Jennie ought to marry the man, and I will consent. I shall be glad to get one eater out of the house, now that you have so improvidently added to the family another daughter."

So Jennie, with her father's consent, was married on her fourteenth birthday. If she had waited four years until she was eighteen, she need not

have had the consent of either parent
or guardian.

How bitter to Mrs. Lex was this lesson of her own powerlessness to prohibit this unfortunate marriage! Her
husband's consent was all that the law
required, and he seemed heedless as to
the great responsibilities and burdens
undertaken by his little fourteen-year-
old daughter, now a wife. Mrs. Lex
thought the State ought to prohibit entirely such early marriages, and that
if the consent of any parent was required, a mother, who knows best the
peril to such a child in marriage, should
be authorized to consent or prohibit as
as well as the father.

Nothing has been said concerning little Rob, who was now about twelve
years old. He was so bright that his
mother thought he might be a physician
some day. She had lately taken a boarder to get money to pay for Rob's school
clothes, but Mr. Lex collected the
money from the boarder, as he was legally entitled to do, (38) and said that
Rob must work in the store; that he had

sent him to school for sixteen weeks already this year, which was all the law required, (39) and the rest of the year he must work. Mrs. Lex was glad enough that the State assumed some control over the child's education, but thought the required time was too short, and she looked and longed for some chance for Rob.

Just then Providence intervened in the form of another relative. Cousin Rose, who lived in Michigan, wrote to Mrs. Lex, saying that if Rob could stay with her for two years to take the place of a son, who had just died, he could go to school and be well cared for. Mr. Lex said he could not go. Mrs. Lex said he ought to, for the sake of the education and the support. She was also afraid Rob would become a loafer, like John and his father, if he stayed with them, while in the Michigan home with her cousin he could learn to be an intelligent, industrious Christian gentleman.

Mary now advised her mother to send Rob anyway. So she asked Mary to take

him to Michigan, but Mr. Lex discovered Rob's absence the day of their
flight and, dashing off, grabbed the boy
at the railway station just before he
boarded the train, and took him home,
meanwhile calling Mary all the names
which he felt she deserved, and telling
her never to come near his family
again. But Mary was not easily scared,
for he had already done his worst to
her. So she tried the plan a few weeks
later and this time successfully.

Mr. Lex knew that his wife had no
legal right to do as she had done, for
his alone was the right to decide upon
the home for his children, but he was
too lazy to exert himself more than to
swear vengeance and threaten that by
next week he would drag Rob home by
the nape of his neck, but he didn't, and
Rob was safe.

Little Dora during this time had not
been a happy baby. Perhaps her
mother's troubles had influenced her
prenatally towards sadness, and perhaps the prevalent irritation at home
had so distressed the mother that Dora
nursed sorrow.

Mr. Lex considered her an unwelcome intruder, and said she cried so much he could not sleep. So he told his wife that Dora must be boarded out somewhere; that, according to the theories of some social reformers, strangers could do better for the child than the mother. He said that Mrs. Lex was too nervous to care well for her and that tomorrow he would take her to a baby farm where she could receive excellent care at a dollar and a half per week. Mrs. Lex pleaded that Dora was only six months old and just beginning her first summer, when it would be dangerous to put her on artificial food. "Just let her stay six months longer." But her husband said, "No;" he knew what was best, and he began to tell her more about his legal rights. Tomorrow the child must be ready for the baby farm. He would take her at noon.

All that sad night Mrs. Lex debated as to what was her duty. Her own lack of control over her children never seemed more unjust. Mr. Lex chose the food and clothes for the children, punished them, medicated them, selected their schools and church, collected wages, selected the burial spot for the dead, decided about the duties of the living, consented to marriage, and now wanted to tear her baby away from her. His guardianship had resulted so injuriously to the other children that she feared the result to the baby. John would never again be strong, and was now a dissolute loafer; brave Mary, driven from home, worked very hard supporting her illegitimate child; Jennie had a drunken husband and a helpless, imbecile baby; and Daisy was in her grave. Only bright Rob was safe. He was still in Michigan. Why could not she go there too? Yes, she would go, she would run away just as soon as her husband left for the store after breakfast.

She got money from Mary for the trip. She started, but found that the Michigan Central train she wanted to take did not leave till twelve o'clock. So she waited. Fatal delay!

She saw her husband enter the station with an officer of the law. She was arrested for the abduction of a child, put into the Harrison Street Police Station, and the baby taken from her. She sent for young Mr. Smith and asked what this horrible charge meant. He explained that she was accused of stealing a child and that,if proved guilty, the maximum punishment was confinement in the county jail one year and a fine of two thousand dollars (40).

Next morning at the hearing before the magistrate, Mrs. Lex admitted that she was trying to take Dora out of the State, but urged that Dora was too young to be put on artificial food and would die if left to the mercies of a baby farmer. Mrs. Lex was much excited and distressed.

Mr. Lex calmly and in correct legal phraseology explained to the Court that he, her father, was the only person who had the right to select the child's home. The mother could not have that privilege.

He said: "Although Dora is young,

courts have frequently gone so far as to 'take nursing infants from the arms of innocent and unexceptionable mothers and place them in the hands of fathers to be reared by adulteresses with whom the fathers were living,' (41) because the father is the child's natural guardian and *prima facie* entitled to custody. (42) A father is entitled to the custody, nurture, and tuition of his child. (43) Even if Mrs. Lex had secured a divorce from me for my own fault and had been granted the custody of the child, this would not give her absolute power over the child. She would not then be in the position of the father. She could act only under the court's direction as any other appointed guardian might. Under the circumstances of divorce, even, our Supreme Court has said when the mother had attempted to take the child out of the State: 'This cannot be tolerated, and must be guarded against.' (44) Many circumstances might curtail the mother's guardianship under order of court, even her remarriage (45). But a father's remarriage has never been held sufficient to divest him of guardianship."

Mr. Lex concluded with this surprising bit of legal information: "Even taking Dora to ride in the street car down to the station contrary to my desire was illegal, as our Supreme Court has decided that a mother has no authority to give permission for a child to have a ride, because the mother is not entitled to disposing power over the person of the child. (46) Of course, I desire to show clearly my own absolute authority over the child, but I also intend to benefit my child. I desire to separate Dora and her mother, as my wife is too emotional to bring up a child."

She seemed emotional then. She was sobbing. The baby too seemed emotional. Though in her father's arms, she was crying vigorously to get to her mother.

To quiet the child the Police Magistrate allowed Mrs. Lex to take her for a few moments, after which he announced his decision. He said:

"The law is clear that Mr. Lex, as the

natural guardian of his child, is entitled to her custody. It does not appear that he is a drunkard or licentious or cruel or a lunatic or unfit to care for her, and even if he was, this Court would have no authority to appoint a new guardian to act in the place of him, the natural guardian. Mrs. Lex is legally guilty of abduction, but there is no desire to make this matter unduly severe nor to subject her to the heavy penalty incurred. As there is no objection, the charge will be changed to that of disorderly conduct and she will be fined one dollar. The child must go to the place selected by the father."

Then Mrs. Lex, instead of being grateful for the small fine, was guilty of contempt of court; for she vowed that she would not let Dora leave her. She said: "That is a cruel, unjust decision. No one would be guilty of so deciding against the beasts of the field or the fowls of the air. Who would take the nursing colt from his mother and give him to his father to rear, saying the father was his natural guardian?

Who would take the brooding mother bird from her little ones?  Better be a beast or bird than a human mother if one desires justice.  Mr. Lex is not the natural guardian of this baby, but I, the mother, am.  Everything this child is, I, under God, made her.  Everything she possesses I gave her."

She continued to speak excitedly, her voice shrill and at times broken with grief:

"What has her father, he whom you call her natural guardian, done for her? Has he given of his own blood and bone and muscle to form her body and to nourish her from day to day, diminishing to just that extent his own strength and vigor? Has he washed and dressed and soothed her not only during the day but through the night as well? Not one of these things has he done, but I, the mother, have done them all. What blasphemous presumption then for him to claim to be the natural guardian! I am the God-ordained, natural guardian of this baby, law or no law.  This is a question of morals, and I shall not let my baby go." She spoke with the courage of despair.

The Judge seemed strangely moved, but Mrs. Lex was a little hysterical and evidently had forgotten that the Harrison Street Police Court was a court not of morals but of law.

Mr. Lex, however, remembered and pointed to the section of the statute giving him sole control. That strengthened the Magistrate's courage, and he directed the bailiffs to take the child. Then ensued a distressing struggle.

Mrs. Lex clung to Dora with all the force of her weak arms, and the baby clung to her, crying with fright. It took some time for the bailiffs to get the child away. They tried to be gentle, and did not want to tear the child into pieces. The Judge was so sympathetic he wiped his eyes and did not fine her for contempt of court. Neither did he fine the baby, who was also vigorously expressing her contempt of court. He was so compassionate! Almost everyone in the room sympathized with Mrs. Lex. The flashily dressed

girls from Custom House Place said
they were glad they had no such husbands; the thieves and pickpockets
waiting their turn, the professional bail-
ors, and the lawyers all showed signs of
emotion. A stony-hearted bootblack,
whose dirty face showed several streaks
where the rare tears had washed away
the grime, piped up indignantly, "I'll
give my pile towards buying the tar for
that father if some one else will come
down with the feathers."

But the law was on the side of Mr.
Lex. Sympathy was on her side, yet
it was not then sufficient for Mrs. Lex.
When at last the bailiffs had torn the
child away the mother fainted. Her
lessons on the legal status of mother
and child were wearing on her.

Mr. Lex then turned the baby over to
the representative of the baby farm,
and took his wife home. For hours she
seemed utterly stunned, but rallied a
little when, next day, after Mr. Lex had
gone to the store, she had a note from
Mary asking her to come to her at
once.

What miracle greeted her eyes? There, in Mary's little cottage with Mary and Mary's child, sat Dora, hungry Dora. What did it mean?

Mary explained that she had sat in the court room veiled in black, had heard the decision, had followed the baby farmer to her place, and soon had appeared in the back yard, where a dozen little ones were wailing in chorus. When the attendant stepped inside, Mary seized Dora and was in the alley before the attendant again appeared, and Dora had been with Mary all night. "Now mother," she said, "this is the safest place in the world for Dora. Father hates me so bitterly he never walks on this street. He keeps the law and makes you miserable. I am going to break it to make you happy, or rather keep a higher law than man's, that is God's. When God gave you food for Dora, it was just as though He commanded from the heavens, 'Feed your child.' You should obey this law. You can easily come over here after breakfast, lunch, and dinner to nurse her; father won't miss you for he will be at the store, and I'll do all the rest if you can trust me." Trust Mary! Yes, she could! And she said such grateful, loving words as warmed poor Mary's heart.

Mary said to her then: "I can appreciate now, since my boy is growing older, how mothers love their children. At first I was too young to realize my disgrace. Then, later, I felt resentment toward my child as though he were the cause of my shame. But now I am learning that he is a blessing instead of a punishment. He is a constant warning and protection against similar wrongdoing on my part, should I ever be tempted, and a constant inspiration to make the most of my life for his sake. I am beginning to dread the ignorance his frequent questions may soon discover and I am beginning my studies just where I dropped them in the grammar school. In this city, as he grows up, he can secure a good education free; and should he want more after the high school is passed, he and I

can work for that. I shall not let him
feel that his birth makes it impossible
for him to succeed, but I shall tell him
of Alexander Hamilton, Henry M.
Stanley and many others who each began life under a cloud as he did and
conquered a place in the world by ability and industry."

Now that Mary had begun, she told
about other thoughts which had been
filling her mind as she worked alone.
"Sometimes I think there was a mis-
take in our Creator's planning and that
both parties to such a wrong should
have a child sent to them, not as a
public curse to bring them deserved
shame, but as a holy blessing to purify
their lives. I have even thought with
pity of the father of my boy, who has
no soft baby arms around his neck, no
tender baby lips pressed to his cheek,
no slender, satin-smooth fingers clinging
to his hand, no soft yielding little body
to press close to him, no child with
awakening mental and spiritual powers
to lift him up. Perhaps if he had he
would not have grown so wicked as he

now is. So, appreciating now the blessing of a little child and knowing how much you must love your baby, I got Dora for you."

It seemed strange that the baby farmer neglected to count her babies for several days, and then neglected to report the loss for several weeks, and that then Mr. Lex was ashamed to tell and Mrs. Lex felt under no obligation to disclose Dora's hiding place. But Mr. Lex knew he was legally right, and that eased his conscience.

Dora got along well in Mary's plain and quiet home. One day Mary said: "Mother, I don't think the laws are fair. Here you have always been good, and yet you have suffered so much through father's having sole authority over your children, while I, who was a bad girl, may God forgive me, have my baby safe, no husband to torture me, no father to torture him and I think I am having a better time than you are. You, the mother of legitimate children, ought to have as much joy in and authority over your children, as I, the mother of an illegitimate child."

"Why, Mary," said Mrs. Lex, "I never thought of making such a comparison between the authority of the mother of an illegitimate child, and of one born in lawful wedlock. But what you say is true."

"Then too," continued Mary, "although I work so hard for small pay; though the neighbors ignore me and churches might not welcome me; yet I am not a slave like you. But if I should now marry my boy's father, the boy would be legitimated and his father would at once have sole control over him. Some weeks ago his father did ask me to marry him. If he had come three years ago, I would gladly have done it, but now I see what he is. I dared not give my boy such a father. I said no. I think he wanted to share in my little home here since his fortune has all been dissipated. So I am more free than you, at least during the first ten years of my child's life. Then the law begins its injustice to me. The father of my child, who never shared in his care during the child's early years, who could not be forced by law to pay over five hundred and fifty dollars for his support during these ten years if the verdict were the highest possible, and who, as an actual fact, never gave one cent for or saw my child, can, when the ten years is past, step up to the Circuit Court with a petition for the custody of my child, alleging that I am unsuitable (47). I am afraid a man judge and men jurors would all decide that I was unsuitable because I was wrong once."

"Don't worry about that," said Mrs. Lex comfortingly, "he won't love your child enough to want him."

"But," said Mary, "then the boy will be nearly ready to earn wages; and whether he loves him or not he might covet his services; and what solace would I have during my old age, I, who have faced scorn and poverty for my child? That is a cruel law; and if it is not changed before my boy is ten, I will move to Colorado, where women are politically equal to men. But I believe there must be changes here.

Things have changed since 1845. Then the father of an illegitimate child might demand possession of the child and upon refusal be no longer liable on the bond he had given for its support."(48)

Mrs. Lex doubted the possibility of ever changing these laws, but admitted that her respect for man-made laws was diminishing.

Mary urged her mother to get a divorce so she could have the custody of her children decreed to her. The daughter perhaps felt more indignant toward her father just then because he had collected some of the money due her from one of her best customers. He had sued the customer and had secured the usual judgment, the Justice stating that the earnings of minors (she was not yet eighteen) belonged to the father.

It was a bitter experience to see her money go into the pockets of the father who had turned her out of the house. As a result she and her child had been reduced to very plain fare for two weeks.

Young Smith said to her later that he wished she had told him and let him fight it. He claimed that her father's expelling her from the home amounted to her emancipation from her liability as a minor. And yet there was a law (49) by which children even when of age were obliged to support poor parents; and if Mr. Lex had not got the money away from her by one method, he might have got it by the other.

But Mary was young and could earn more money, and her trouble did not make her meek but inspired her to retaliation. She asked Smith about the divorce so her mother could have control of the children. He said Mrs. Lex had no legal ground. Mr. Lex was not licentious, nor a drunkard, nor a bigamist, nor had he struck her, nor attempted to take her life, nor deserted her; (50) nor could she even bring a suit for separate maintenance as they were still under the same roof.

So, with the hope of securing the control of her children gone, Mrs. Lex continued to bear her burden.

Some weeks after this either the unexplained loss of Dora, never confided to his wife, or some hereditary trait caused Mr. Lex to act so strangely and wildly that the neighbors thought he was insane, and were so frightened that they had him sent to the detention hospital.

At the trial of his sanity among other eccentric freaks which were related were his actions toward the members of his family in exercising his authority as guardian entitled to the custody and control of the children.

Mr. Lex defended himself with considerable skill. He admitted that recently he might have done a few things to the neighbors which the law did not allow, but said that in his relations with his family he had followed the law strictly. He read from the statute and cited one case after another from the reports.

Lawyer Smith spoke on the other side and very severely. He said: "No man who was sane would think of enforcing those laws so strictly. They have come down to us from an ancient time when conditions were different, when women and children were kept behind walls for their own protection from bands of roving destroyers, when the mailed and armed father was the chivalrous guardian of all. To him, the strongest, the owner of property, the warrior, was full authority given.

"At the time these laws originated no one contemplated that a father might be as improvident and as incapable as was Mr. Lex, nor could anyone have foreseen the changing position of woman and her gradual development into a stronger person more fitted to rule her children then was the shut-in woman of centuries ago, who never went beyond her walls, could not read, and thought but as a child. These laws, under which Mr. Lex acted, originated long ago and have not been changed to keep up with the spirit of the times. Many do not realize their possible injustice because many husbands and fathers have kind hearts and never think of living up so strictly to their ancient legal rights. If only for one day every husband in Illinois should do these things done by Mr. Lex, there would be such a rebellion against their authority by all women that we men would find less complaining in Tophet." At this point several married men present nodded significantly.

Mr. Smith said, in concluding his speech:

"In the coming Legislature, of which I expect to be a member, I shall do all that one man can to repeal such unjust, cruel laws and to bring equal rights to women. So long as one woman can be made to endure such wrongs as Mrs. Lex has borne, all under the sanction of law, then the happiness, the liberties, the rights of all women are endangered. I shall want these laws changed for man's sake too. Uncontrolled, irresponsible power is often the greatest curse to him who wields it. Let us, as men, put away from us such opportunities for cruelty and injustice and give mothers equal power with fathers. But with these laws as they now stand no sane man should think of enforcing them. That Mr. Lex did is sure proof of his insanity."

Mr. Smith did speak sharply, but the jury seemed to agree with him and soon returned with a verdict that Mr. Lex was insane, and the Judge had an order entered for his commitment to the asylum.

The Judge said: "In view of all the facts the verdict of the jury seems a right one; but my young friend Mr, Smith is a little too severe in his objections to the law relating to the father's unlimited control over the child. It is true the State makes the father guardian and custodian. Yet the State retains some supervision, and there are many statutes for the child's benefit limiting paternal control, as it existed under the old common law; for example: the statutes forbidding certain employments (51), forbidding the giving of tobacco to minors under sixteen years of age (52), compelling sixteen weeks' school attendance (53), prescribing penalties for wilful cruelty (54), for endangering life or health (55) and for aban-

donment of a child under one year of age
(56). In addition to this restriction on
paternal authority, the father cannot ap-
prentice the child without the mother's
consent (57), nor remove the family
from the homestead without her consent unless he has furnished another
home (58), nor dispose of the custody
and tuition of his child by will if the
mother be suitable and desirous of act-
ing as guardian (59), nor deprive her
of her child's custody if he is a lunatic.
(60).

"It is being more generally recognized
that the mother is intimately connected
with the child. In a recent decision,
Allaire vs. St. Luke's Hospital, 76 Ill.,
App. 540, the Court held that the child
before birth was a part of the mother
and not a separate entity. Public
opinion approves considering a very
young child a mother's special care,
and many fathers, recognizing the
mother's peculiar fitness for deciding
matters affecting a child's welfare,
never think of exercising their own
legal authority. Even when there is a

contest between the parents, and judges have the discretion to award the custody of a child to whichever of the parents seems best fitted for the duty, judges often select the mother for very young children. Perhaps the discretion of men judges may occasionally lead them to grant the custody to unworthy fathers, but with the thought before them that the child's good is the end sought, judges have often felt free to recognize the mother's ability. In view of this progress there is ground for hope of greater extension of mothers' privileges."

The Judge seemed much interested in this matter, although it did not directly affect the case at bar. He went on to say, "Some of the annoyances which Mr. Lex has caused were the fault, not of law, but of industrial conditions. "If Mrs. Lex had been in the store receiving pay for groceries, she would have had some ready cash with which to purchase supplies and to exercise her judgment; and if at the same time Mr. Lex had been forced to do housework at home and care for the children with no salary, he might have been obliged to submit to her judgment in many things. But it was not so. Mrs. Lex, like nine-tenths of the women in the United States, did all her own work and had no wages. Without money she was powerless to carry out her own wishes. Women must have some financial resources before they can ever stand equal to their husbands in the control of their children. If women continued financially dependent, they would still be inferior.

"As to the specific acts done by Mr. Lex, none of them were criminal, none of them would be sufficient ground for a divorce. Apparently they were all sanctioned by the law of this State. All men must be ashamed that the severest portions of these laws have not been changed and the burdens of the so-called weaker sex lightened. Mr. Lex has been law-abiding to the point of atrocity and insanity. He must go to the asylum."

This was a blow to Mr. Lex. To be

incarcerated in an asylum for supporting too well the constitutions and law he had sworn to uphold when he was admitted to the bar, dazed him.

When Mrs. Lex visited him at the asylum, he seemed scarcely to know her. The physicians told her he had been irresponsible for a long time. Then she forgave him all, and remembered him only as the tender, loving husband he had been years ago.

In a few months he died and the only mourner was the woman whom he had so cruelly wronged.

Mrs. Lex now carries on the store. John, who is now trying to form better habits, is her helper. Rob is now back in Chicago finishing at the high school. Jennie has been divorced from her drunken husband, the Court allowing her to keep the two children, one of whom is an imbecile. She is embroidering for the Woman's Exchange, and is helping Mary sew. Mary took them all into her little cottage. Mrs. Lex, John, Rob, Jennie and her two children, little Dora and Mary's child, now comprise the family.

Timid Mrs. Lex is now less timid; and one day she startled Mary by saying: "I have thought many times about what the Judge and Mr. Smith said at your father's trial. The laws ought to be changed, but I don't believe they will be until the majority of women, seeing what sad things are possible under the law, work for and demand changes. I fear that men alone will not so clearly see the need. Some men surely know about these injustices; for they are judges, jurors, police magistrates, constables, and hear many of these sad cases; and yet they seem too fond of power or too indifferent to our sufferings to make the necessary changes. To the individual woman, as to me, they often try to be merciful, but the law forbids their being just, and they forget that the law unchanged makes it possible for every woman to suffer as I have suffered. Very few women have learned in the bitter school of experience as I have about the legal status of mother and child. Yet, as all women may become mothers, there is a chance that all may suffer. I do not pretend that my wrongs are the only ones. I know that wage-earning women receive less than men for the same work; that many schools of technical training are still closed to women; that custom keeps women away from many remunerative employments, and that home women do not have wages.

"All this injustice should cease. Yet my sufferings have been from another cause. The law gave me no control over the children who were part of my being, when I, next to God, was their creator. So I think especially of this. Nor am I the only woman who has suffered from this cause. Many women have confided their troubles to me, and I find that in many homes mothers and children are being wronged through the fathers' ignorant yet unlimited, domineering control. While my wrongs were greater than some of theirs, yet I was in no way better fitted to endure them. I was only an average woman. If I had been unusually brilliant and had been able to win fame and fortune with pen or brush or voice, I could have helped myself and my children better. Or if I had been very beautiful, admiring men, jurors and judges, might have strained a point in my favor, especially if I had also been fascinatingly wicked.

"My love for my children, which grew to be the strongest force in my life, even that did not make me noticeably different from the majority of mothers. I did no more for my children, sacrificed no more for them, than the majority of mothers do or are willing to do. I was only an average mother. So I believe other women, all mothers, would feel as I do and be glad to help. Cousin Jane and Cousin Rose, who knew our troubles, felt for us; and other women, if they should learn about such things, surely would not be so indifferent as men have been, but would unite with us in securing permanent changes."

"But, mother," said Mary, "what can you and I do? We work so hard every day that we have no time left for any thing else. Would I, the mother of an illegitimate child, be listened to in any woman's club or church? I could not even use correct English, I, a mother before I was fifteen years old. I would not be good enough to speak from the lofty standpoint of unassailed virtue, nor would I be bad enough to make an interesting, horrid example. My story, ruined before fifteen, supported child ever since, is too commonplace to win attention. If I could relate a thrilling tale of vice, of years of evil, I would still be welcomed cordially to the beautifully-kept reformatories of the most charitable, and the kind ladies of the Board might enjoy hearing my experiences. They would feel very complacent about having furnished me a comfortable place in which to die, and would be so busy at that that they would have little time to change laws. So I am not good enough and not bad enough to be an attractive or popular speaker. And you, timid little mother, with your weak voice and bent back and shabby clothes, would they listen to

you, especially if some one should e
that you were once imprisoned for
stealing a child and found guilty? No,
women generally would not listen. If
they thought ⌐y were to discover
such sad things, they would shut their
ey ﾝ and stick their fingers in their ears
for fear their own comfort with indul-
gent husbands would be disturbed."

"You are bitter, Mary," said Mrs.
Lex, "but women's hearts are kind.
You and I cannot speak or go to meet-
ings, that is true; we have no powerful
friends, no gift of language; but the
time is coming when women of education and wealth and leisure and social
prestige will take up the cause of
women, other women, all women. Just
now some of them are more interested,
apparently, in 'society,' clubs, Hindoo
philosophy, whist, or the heathen or
drunkards, criminals or incompetents.
They are busy studying, entertaining,
reforming, curing. They are only sleep-
ing on the question of women's needs.
They will wake up soon to the injustice
possible to all their sex, and will begin

a work greater than that of charity or of philanthropy. They will begin to secure justice. Then mothers will be equal to fathers."

# TABLE OF LEGAL CITATIONS.

Abbreviations used:

R. S.—Revised Statutes of Illinois.

162 Ill., 361.—Volume 162 of the Illinois Supreme Court Reports, page 361.

Ill. App.—Illinois Appellate Court Reports.

Am. & Eng. Ency. of Law.—American and English Encyclopedia of Law. First Edition.

1. Chap. 68, Sec. 9, R. S.
2. Chap. 148, Sec. 1, R. S.
3. Hyman vs. Harding, 162 Ill., 361.
4. Garfield vs. Scott, 33 Ill. App., 319.
5. Smith vs. Slocum, 62 Ill., 358.
6. Barrett vs. Riley, 42 Ill. App., 260.
   McMahon vs. Sankey, 133 Ill., 648.
   Magee vs. Magee, 65 Ill., 255.
7. Capek vs. Kropik, 129 Ill., 515,
   Koster vs. Miller, 149 Ill., 200.
8. Bradley vs. Sattler, 54 Ill. App., 506, 156 Ill., 608.
9. Temple vs. Freed, 21 Ill. App., 239.
10. Woodyatt vs. Connell, 38 Ill. App., 481, Chap. 68, Sec. 16, R. S.
11. Chap. 38, Sec. 237, R. S.
12. Chap. 38, Sec. 237, R. S.
13. Chap. 17, Sec. 8, R. S.

14. Chap. 17, Sec. 14, R. S.
15. Vetten vs. Wallace,39 Ill. App.,390.
16. Glidden vs. Nelson,15 Ill. App.,297.
17. Chap. 68, Sec. 24, R. S.
18 Ford vs. McKay, 55 Ill., 119.
19. 21 Am. & Eng. Ency. of Law, 1009 and 1016.
20. Grable vs. Margrave, 4 Ill., 372.
21. White vs. Murtlandt, 71 Ill., 251. Garretson vs. Becker, 52 Ill. App., 255. Bayles vs. Burgard, 48 Ill. App., 371.
22. Ball vs. Bruce, 21 Ill., 162.
23. Hobson vs. Fullerton, 4 Ill. App., 284.
24. Scharf vs. People, 134 Ill., 240, and cases there cited. Maynard vs. People, 135 Ill., 430. Coleman vs. Frum, 4 Ill., 378.
25. Hudson vs. King Brothers, 23 Ill. App., 122.
26. Argument of E. H. & N. E. Gary in Walcott vs. Hoffman, 30 Ill. App., 79.
27. Cole vs. Bentley, 26 Ill. App., 260. Walcott vs. Hoffman, 30 Ill. App., 77.

28. Hudson vs. King Brothers, 23 Ill. App., 122. Glaubensklee vs. Lew, 29 Ill. App,, 413.
29. Chap. 68, Sec. 15, R. S.
30. Chap. 52, Sec. 13, R. S.
31. John V. Farwell & Co. vs. Martin, 65 Ill. App., 57.
32. Clinton vs. Kidwell, 82 Ill., 427.
33. Barrett vs. Riley, 42 Ill., App., 260. Tyler vs. Sanborn, 128 Ill., 144.
34. Temple vs. Freed, 21 Ill. App., 239.
35. Bradley vs. Sattler, 156 Ill., 608.
36. Chap, 68, Sec. 16, R. S.
37. Chap. 89, Sec. 13, R. S.
38. Parker vs. Parker, 52 Ill. App., 333.
39. Chap. 122, Sec. 313, R. S.
40. Chap. 38, Sec. 2, R. S.
41. Miner vs. Miner, 11 Ill., 50. 17 Am. & Eng. Ency. of Law, 364.
42. People vs. Porter, 23 Ill. App., 196; Umlauf vs. Umlauf, 128 Ill., 383. Perry vs. Carmichael, 95 Ill., 530; Wright vs. Bennett, 7 Ill., 587; Bedford vs. Bedford, 136 Ill., 361.
43. Chap. 64, Sec. 4, R. S.

44. Miner vs. Miner, 11 Ill., 51.
45. 17 Am. & Eng. Ency. of L., 365; See opinion of Chief Justice Breese in Miner vs. Miner, 11 Ill., 43.
46. Pierce vs. Millay, 62 Ill., 134.
47. Chap. 17, Sec. 13, R. S.
48. Wright vs. Bennett, 7 Ill., 587.
49. Chap. 107, Sec. 2, R. S.
50. Chap. 40, Sec. 1, R. S.
51. Chap. 38, Sec. 42, a. R. S.
52. Chap. 38, Sec. 42, f. R. S.
53. Chap. 122, Sec. 313, R. S.
54. Chap. 38, Sec. 53, R. S.
55. Chap. 38, Sec. 42, d. R. S.
56. Chap. 38, Sec. 42, h. R. S.
57. Chap. 9, Sec. 2, R. S.
58. Chap. 68, Sec. 16, R. S.
59. Chap. 64, Sec. 5, R. S.
60. Chap. 86, Sec. 5, R. S.
61. Stolz vs. Doering, 112 Ill., 239.